The Upcoming Fight Between Tesla, Chinese EV Makers, and Other EV Makers in America

Introduction:

The electric vehicle (EV) market is undergoing a significant transformation, with major players like Tesla facing increasing competition from Chinese EV makers and other international and domestic automakers.

This burgeoning competition is expected to drive innovation, improve infrastructure, and ultimately benefit EV buyers and owners.

As these companies vie for dominance in the American market, the landscape of EV technology, infrastructure, and consumer experience will evolve in profound ways.

Historical Context and Current Landscape

Tesla's Dominance:

Since its inception in 2003, Tesla has become synonymous with electric vehicles.

Under the leadership of Elon Musk, Tesla has revolutionized the automotive industry with its cutting-edge technology, sleek designs, and commitment to sustainability.

Tesla's flagship models, such as the Model S, Model 3, Model X, and Model Y, have set new standards in performance, range, and luxury.

The company's robust Supercharger network has alleviated range anxiety for many consumers, making EV ownership more practical.

Tesla's vertical integration strategy, encompassing everything from battery production to software development, has given it a significant competitive edge.

The company's focus on innovation is evident in its developments in autonomous driving technology, energy storage solutions, and the launch of the Cybertruck and the Tesla Semi.

Rise of Chinese EV Makers:

China's aggressive push towards electric mobility has led to the emergence of several formidable EV manufacturers, including NIO, BYD, and Xpeng.

Supported by substantial government subsidies and a vast domestic market, these companies have rapidly advanced their technology and scaled production.

NIO, known for its premium electric SUVs and innovative battery-swapping technology, has positioned itself as a luxury brand.

BYD, backed by Warren Buffett's Berkshire Hathaway, is a leader in battery technology and electric buses.

Xpeng, with its focus on autonomous driving and smart features, targets tech-savvy consumers.

These companies are now eyeing the American market, bringing competitive pricing, innovative technology, and a fresh perspective on EV ownership.

Other Key Players:

Traditional automakers like General Motors (GM), Ford, and Volkswagen are also making significant strides in the EV space.

GM's Ultium battery technology and ambitious plans to phase out internal combustion engines by 2035 highlight its commitment to electrification.

Ford's Mustang Mach-E and the electric F-150 Lightning have generated substantial interest.

Volkswagen's ID.4 and other models represents its serious foray into the EV market.

Additionally, startups like Rivian and Lucid Motors are entering the scene with high-performance electric trucks and luxury sedans, respectively.

These new entrants add to the competitive dynamics and push established players to innovate further.

Competitive Strategies

Tesla's Approach:

Tesla continues to leverage its first-mover advantage by expanding its product lineup and enhancing its technological capabilities.

The company's focus on software, particularly its Full Self-Driving (FSD) technology, aims to offer unmatched convenience and safety features.

Tesla's Autopilot and FSD packages promise a future where cars can drive themselves, revolutionizing personal transportation.

The expansion of Gigafactories worldwide aims to increase production capacity and reduce costs.

Tesla's energy products, such as solar panels and Powerwall home batteries, complement its EV offerings, promoting a holistic approach to sustainable energy.

Strategies of Chinese EV Makers:

Chinese EV makers are leveraging their cost advantages and technological innovations to penetrate the U.S. market.

They offer vehicles at competitive prices, appealing to cost-conscious consumers.

Companies like NIO emphasize customer service innovations, such as battery swapping, which can significantly reduce charging times.

BYD's expertise in battery technology and Xpeng's focus on smart features and autonomous driving give them a competitive edge.

Strategic partnerships with American companies for production and distribution could help Chinese manufacturers establish a foothold in the market.

Collaborations with tech firms for autonomous driving and infotainment systems can enhance their offerings.

Strategies of Traditional and Other New EV Makers:

Traditional automakers are leveraging their brand loyalty, extensive dealership networks, and production capabilities to transition to electric mobility.

GM's investment in Ultium battery technology and Ford's electric F-150 Lightning are part of broader strategies to capture market share.

Startups like Rivian and Lucid Motors are focusing on niche markets with high-performance and luxury electric vehicles.

Rivian's electric trucks and SUVs target adventure enthusiasts, while Lucid's luxury sedans aim to compete with Tesla's high-end models.

Market Dynamics and Consumer Impact

Pricing and Affordability:

One of the most significant impacts of the competition between Tesla, Chinese EV makers, and other players will be on pricing.

Increased competition is likely to drive down the cost of EVs, making them more accessible to a broader range of consumers.

This could accelerate the transition from internal combustion engine vehicles to EVs, contributing to environmental goals and reducing dependence on fossil fuels.

Technological Advancements:

The rivalry is also expected to spur technological advancements.

Both Tesla and Chinese manufacturers are heavily invested in research and development.

This competition could lead to faster improvements in battery technology, extending the range and lifespan of EVs.

Additionally, advancements in autonomous driving technology could make self-driving cars a reality sooner than anticipated.

Infrastructure Development

Charging Networks:

The expansion of charging infrastructure is critical for the widespread adoption of EVs.

Tesla's Supercharger network is currently one of the most extensive and reliable.

However, the entry of Chinese EV makers will necessitate the development of compatible charging solutions.

This could lead to increased investment in public charging stations, benefiting all EV owners by reducing charging times and increasing convenience.

Battery Technology:

Battery technology is at the heart of the EV revolution. Tesla's Gigafactories have set a high standard for battery production.

In response, Chinese manufacturers are likely to establish their own production facilities in the U.S., fostering competition and innovation.

This could result in more efficient, longer-lasting, and cheaper batteries, further driving down the cost of EVs and enhancing their appeal.

Government Policies and Regulations

U.S. Policies:

The U.S. government has shown strong support for the EV industry through tax incentives, subsidies, and regulatory measures aimed at reducing carbon emissions.

The entry of Chinese EV makers could prompt policymakers to introduce additional measures to support domestic manufacturers.

However, the competition could also encourage a more balanced approach, ensuring that consumers benefit from a wider range of choices and lower prices.

International Trade Considerations:

The geopolitical dynamics between the U.S. and China could influence the competition in the EV market.

Trade policies, tariffs, and regulatory standards will play a crucial role in shaping the strategies of Chinese EV makers.

Navigating these complexities will be essential for both Tesla and its Chinese competitors as they vie for market share.

Who Will Eventually Win?

Tesla's Strengths:

Tesla's established brand, loyal customer base, and technological prowess give it a strong position in the American market.

The company's continuous innovation and expansion plans will help maintain its competitive edge.

Tesla's ability to integrate hardware and software seamlessly offers a unique value proposition that is hard to replicate.

Chinese EV Makers' Potential:

Chinese EV manufacturers bring significant strengths to the table, including cost-effective production, rapid innovation cycles, and a willingness to adopt new business models.

Their success in the U.S. market will depend on their ability to adapt to local consumer preferences and regulatory requirements.

If they can overcome these challenges, they could capture a substantial market share, particularly in the budget and mid-range segments.

Traditional Automakers and New Entrants:

Traditional automakers like GM and Ford, with their extensive resources and brand recognition, are well-positioned to compete in the EV market.

Their ability to leverage existing infrastructure and customer loyalty will be crucial.

Startups like Rivian and Lucid Motors, with their niche focus and innovative approaches, will add diversity and choice to the market, appealing to specific segments.

Implications for EV Infrastructure

Charging Infrastructure:

The increased presence of Chinese EV makers will likely lead to a more diverse and extensive charging infrastructure.

Collaboration and competition between different stakeholders could result in a more robust network of charging stations, benefiting all EV users.

Innovations such as faster charging technologies and battery swapping stations could become more prevalent.

Standardization and Interoperability:

As more players enter the EV market, the need for standardization and interoperability of charging solutions will become more pressing.

Efforts to create universal standards for charging connectors and payment systems could simplify the experience for consumers, making EV ownership more convenient and attractive.

Benefits for EV Buyers and Owners

Lower Prices:

Increased competition between Tesla, Chinese EV makers, and other players will likely drive down the cost of EVs, making them more affordable for a broader range of consumers.

This price reduction will lower the barrier to entry for many potential buyers, accelerating the transition to electric mobility.

Enhanced Features and Options:

The rivalry will also lead to a greater variety of models and features available to consumers.

Buyers will have more options in terms of design, performance, and price, allowing them to choose an EV that best suits their needs and preferences.

This diversity will make the EV market more dynamic and consumer-friendly.

Better After-Sales Service:

Chinese EV makers, known for their customer-centric approaches, may introduce innovative after-sales services in the U.S. market.

Enhanced warranties, flexible financing options, and efficient maintenance services could improve the overall ownership experience for EV buyers.

Challenges and Considerations

Market Penetration:

For Chinese EV makers and new entrants to succeed, they must navigate the complexities of market penetration in the U.S.

This involves understanding local consumer preferences, establishing a reliable distribution network, and complying with stringent regulatory standards.

Consumer Trust and Brand Loyalty:

Building consumer trust and brand loyalty is crucial for any new entrant in the automotive market.

Chinese EV makers and startups will need to invest in marketing, customer service, and product quality to compete with established brands like Tesla and traditional automakers.

Geopolitical and Trade Issues:

The geopolitical landscape and trade policies between the U.S. and China could impact the EV market.

Tariffs, trade restrictions, and regulatory differences may pose challenges for Chinese EV makers entering the U.S. market.

Navigating these complexities will require strategic planning and diplomacy.

The Future of EV Technology

Autonomous Driving:

Autonomous driving technology is one of the most promising areas of development in the EV industry.

Tesla's Full Self-Driving (FSD) system is at the forefront, but Chinese EV makers and other competitors are also making significant strides.

Advances in AI, sensor technology, and regulatory approval will be critical to the widespread adoption of autonomous vehicles.

Battery Innovations:

Battery technology continues to evolve, with advancements in energy density, charging speed, and cost reduction.

Solid-state batteries, which promise higher energy density and safety, are on the horizon.

Companies like Tesla, BYD, and others are investing heavily in battery research to maintain their competitive edge.

Vehicle-to-Grid Technology:

Vehicle-to-Grid (V2G) technology allows EVs to feed electricity back into the grid, providing a decentralized energy storage solution.

This technology can enhance grid stability, support renewable energy integration, and offer financial incentives to EV owners.

As more EVs hit the road, V2G could become a significant component of the energy ecosystem.

Conclusion

The upcoming fight between Tesla, Chinese EV makers, and other players in the American market will be a transformative period for the EV industry.

This competition will drive innovation, lower prices, and expand the EV infrastructure, ultimately benefiting consumers and accelerating the adoption of electric vehicles.

While Tesla holds a strong position, the entry of Chinese manufacturers and the efforts of traditional automakers and startups will inject new energy and diversity into the market.

The ultimate winners will be the consumers, who will enjoy more choices, better technology, and an improved EV ownership experience.

As these companies compete for market share, the resulting advancements in EV technology and infrastructure will pave the way for a more sustainable and efficient transportation system.

The future of the EV market in America is bright, with the promise of exciting developments and opportunities for all stakeholders involved.

Please use the next few pages for your notes and debates.

www.ingramcontent.com/pod-product-compliance
Lightning Source LLC
Chambersburg PA
CBHW070120230526
45472CB00004B/1339